# Whale Watching in Iceland

ÁSBJÖRN BJÖRGVINSSON & HELMUT LUGMAYR
Illustration: Martin Camm
Translation by: Jón Skaptason

JPV ÚTGÁFA

*Whale Watching in Iceland*

Designer & Creative editor: Jón Ásgeir í Aðaldal
Illustrator: © Martin Camm
Photographers: Friðþjófur Helgason, Heimir Harðarson,
Jóhann Óli Hilmarsson, Jón Ásgeir í Aðaldal, Mark Carwardine
Printed by:

JPV PUBLISHERS · Reykjavík · 2002



ISBN: 9979-761-55-5

# Contents

# Introduction

When I was a boy living in Flateyri in the West Fjords of Iceland, I remember being seriously frightened in an incident involving a few porpoises. They came leaping and frolicking in the direction of the rowboat that I and my friends used to go fishing in when the weather was good.

The cry went up, "whales, whales!" when the porpoises came dancing over the waves. In our minds, whales were dangerous beasts that could easily flip over a small boat and swallow the crew in a single gulp. There was no option but to head for shore, rowing for dear life, when these creatures were sighted on our fishing trips. We were totally ignorant about whales, and to us the harmless porpoises were menacing leviathans.

In my recollection, I always associated whales with the arrival of the small minke-whaling boats in Flateyri harbour. There was always great bustle and excitement when they came chugging into the fjord with one or two minke whales hanging over the side. The whales were hauled ashore and butchered on the pier with huge knives. The townspeople would often bring wheelbarrows to take back the meat they bought, and for the next few months everyone would have whale meat on their tables, cooked in the traditional manner after soaking in milk to get rid of the fatty taste.

In Iceland, minke whaling was stopped in 1984 and all whaling ceased in 1989. Most of the whale stocks are recovering from centuries of excessive hunting, and are now being utilised in different and more environmentally friendly ways than before. Whale watching has increased dramatically in recent years, and is now one of the mainstays of the Icelandic tourist industry.

To my mind, whale watching is communion with nature at its best. To sail on the open sea or in the fjords, in calm weather or blustery winds, rain or shine, with a grand view of the mountains and birds all around awakens a sense of pure elation and awe at nature and life itself.

We hope this little book will not only inspire you, dear reader, to see for yourself, but also that it will give added meaning to your excursion and deepen your understanding of those magnificent animals, the whales.

*Ásbjörn Björgvinsson and Helmut Lugmayr.*

# The Origin of Whales

*The oldest whale skeletons are approximately 50 million years old.*

Whales have adapted entirely to life in the sea, unlike other marine mammals, such as seals or sea lions, which are to some extent dependent on land. If whales are washed ashore, for any reason, they will suffocate from the pressure of their own weight. The evolution of whales involved massive changes in skeletal and physical structure and the development of organs which are unique to whales.

The oldest known skeletons of whales (found in Britain, Egypt, Western Africa and New Zealand) are 50 million years old; these skeletons are remains of animals that were obviously aquatic. It is not quite clear what species of land mammals the whales evolved from, but they appear to be most closely related to an ungulate species of the order of *Condylarthra,* an animal about the size of a dog or horse, and widespread in the world following the extinction of the dinosaurs.

There are now some 78–80 identified species of whale. Some whales live only in the open oceans, and others are extremely wary and rarely observed, so that not much is known about them.

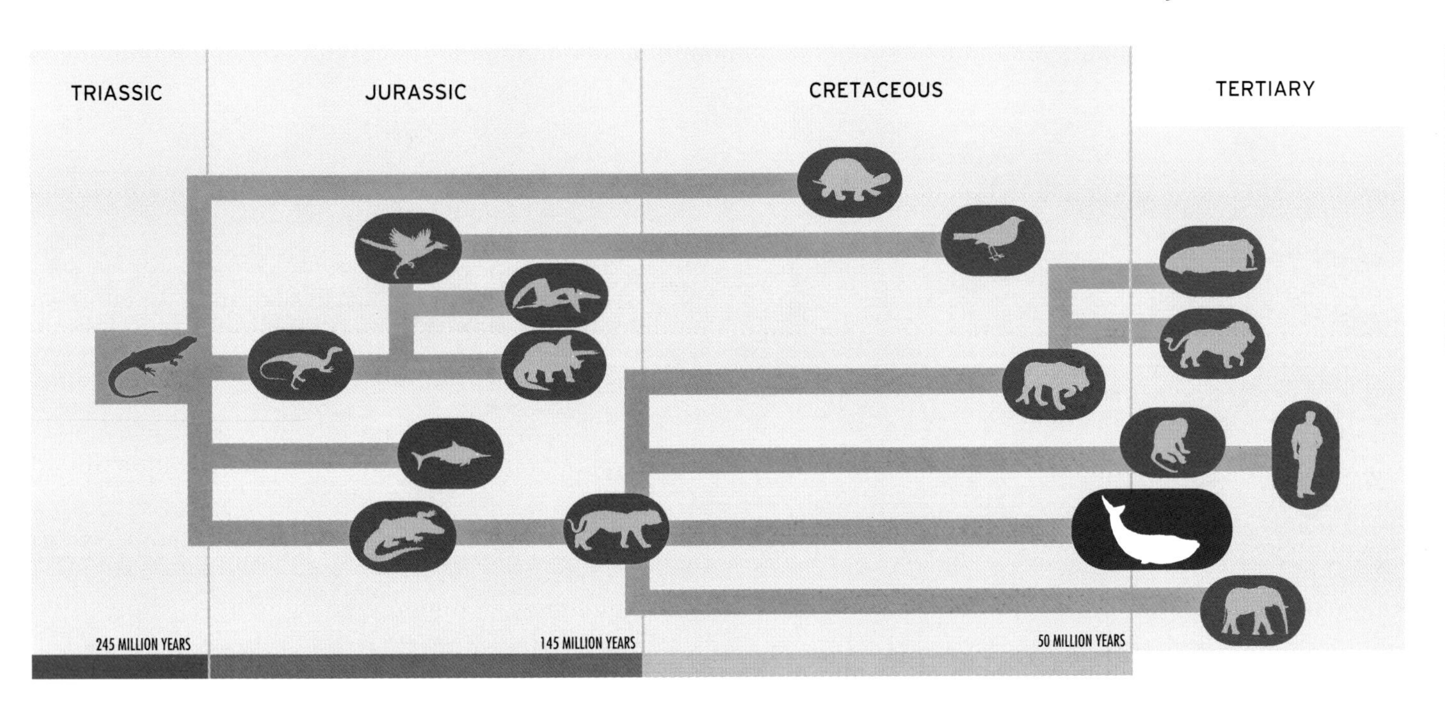
TRIASSIC
JURASSIC
CRETACEOUS
TERTIARY
245 MILLION YEARS
145 MILLION YEARS
50 MILLION YEARS

# Characteristics



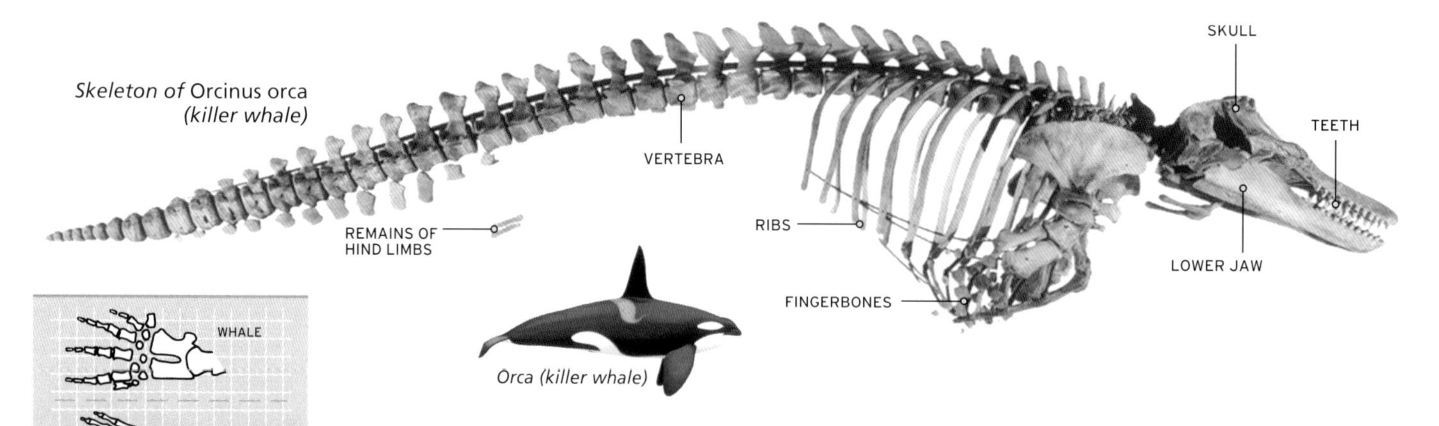


*Skeleton of* Orcinus orca *(killer whale)*

*Orca (killer whale)*

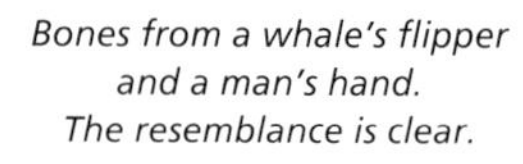

*Bones from a whale's flipper and a man's hand. The resemblance is clear.*

**SKELETAL STRUCTURE AND PHYSIOLOGY**

The most characteristic feature of whales, like other aquatic animals, is their streamlined shape. The cervical vertebra are virtually fused together in almost all the species. The head shape is characterised by unusually large protruding jaws, and the nostrils, which take the form of either one or two blowholes, are located on top of the head.

**Extremities**

The extremities are unobtrusive, enhancing the sleekness of the body. Forward extremities have developed into flippers with no visible fingers, and hindquarters are missing altogether. Of the pelvic structure, only a few undeveloped bones remain.

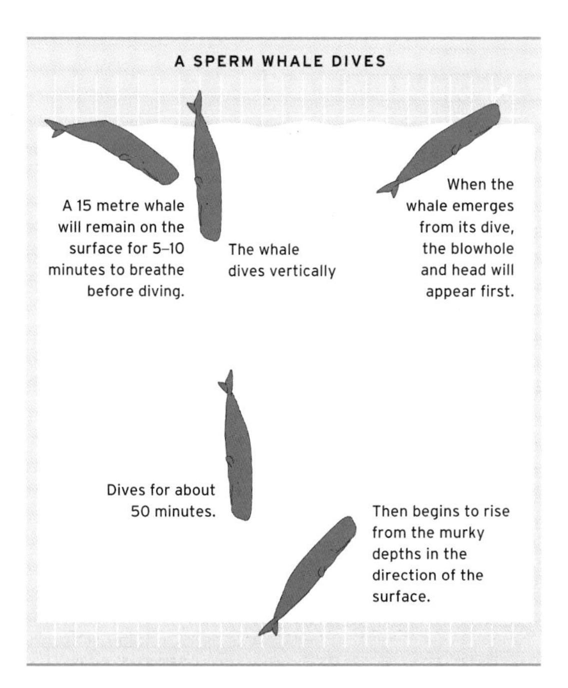


## Movement

The whale's principal means of propulsion is the tail fin, or flukes, which, unlike fishtails, are horizontal. Using their flukes, the largest baleen whales and orcas can reach speeds of up to 50 kilometres per hour. There are no bones in cetacean tail fins or dorsal fins, only fibre, cartilage and fat.

Animals this size would hardly be able to move on land, but in the water they are virtually weightless.

## RESPIRATORY SYSTEM AND CIRCULATION

Whales must be able to hold their breath for as long as possible. This is obviously of vital importance when diving deep in search of food, or "sounding".

How long and how deep whales dive depends on the type of food they are after. The record probably belongs to the sperm whales, which can dive to a depth of two kilometres in one hour!

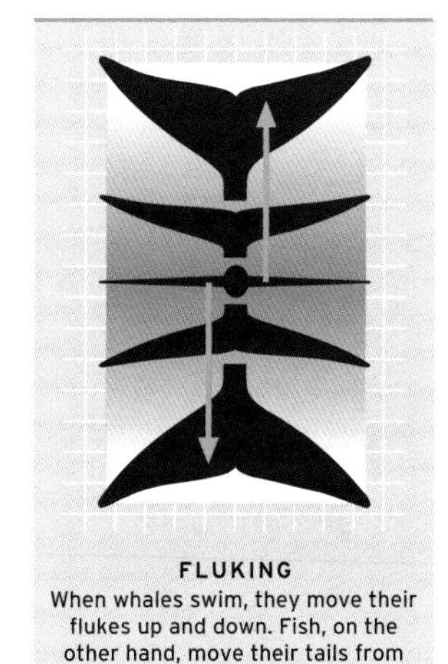

**FLUKING**
When whales swim, they move their flukes up and down. Fish, on the other hand, move their tails from side to side.

# Characteristics

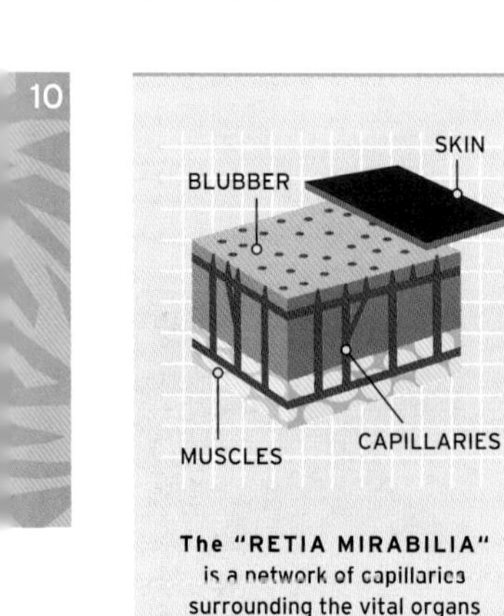


**The "RETIA MIRABILIA"** is a network of capillaries surrounding the vital organs of whales.

**Lungs**

Oddly enough, whales' lungs are no larger, proportionally, than the lungs of land mammals. The lungs are located lengthways beneath the spine and maintain stability in the sea, much like the swim bladder in fish, while the stomach serves as ballast. The diaphragm between the stomach and lungs is unusually powerful, enabling whales to breathe in an incredible amount of air in a very short time. Fin whales can inhale and exhale about 1,500 litres of air in two seconds, exchanging some 90% of the air in their lungs. In comparison, the corresponding figures for humans is half a litre of air every four seconds, which represents some 20% of the air in the lungs.

**The "Wonder Net" *(Retia mirabilia)***

The respiratory system and blood circulation of whales must ensure that all organs receive sufficient oxygen during long dives. To accomplish this, whales are endowed with a special organ known as the *retia mirabilia,* which is Latin for "wonder net". The retia is a network of capillaries surrounding the vital organs of whales. The capillaries are a sort cross between arteries and veins capable of storing oxygenrich blood. Almost all the oxygen that enters the lungs on inhalation enters the bloodstream and is not stored only in the red corpuscles, but also in the myoglobin, which explains the dark colour of whale meat. In other words, it is not the quantity of air in the lungs that allows whales to remain submerged for so long, but rather their ability to squeeze every bit of oxygen from the air they inhale and store it in their muscles and retia.

**Regulation of Body Temperature**

In addition to storing oxygen, the retia serves the purpose of equalising body temperature and pressure. There is no burning of oxygen in the retia, but it performs other useful functions, e.g. by cooling the blood and preventing overheating, or by releasing heat and thereby protecting organs from excessive cooling when sounding.

The outer layer of fat, the blubber, provides the main insulation for the body; the thickness varies, depending on species, from 5 to 50 centimetres. The limited circulation of blood to the outer layers of the body also reduces heat loss. The insulating properties of whale blubber are so great that when a dead whale

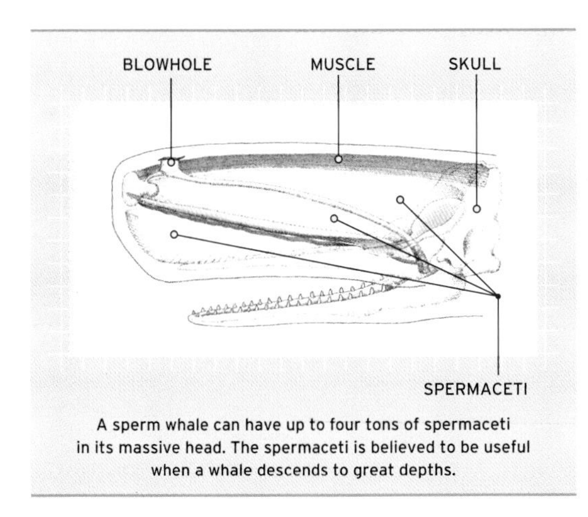


A sperm whale can have up to four tons of spermaceti in its massive head. The spermaceti is believed to be useful when a whale descends to great depths.

is washed ashore and the carcass begins to decompose, the temperature within rises rapidly from 37° to 50–60°.

**Pressurisation**

The tremendous change in pressure when whales sound causes extreme physical stress. The layer of blubber contributes to pressurisation, but in addition some species of whale have a store of liquid fat in the head known as "spermaceti". The spermaceti is believed to be useful when a whale descends to great depths. A sperm whale can have up to four tons of spermaceti in its massive head. This makes it possible for them to dive to depths of up to two kilometres, where they need to withstand a pressure of about 200 bar, which corresponds to two thousand tons per square metre. Vital blood vessels, such as the arteries leading to the brain, lie within the spine, which protects them from collapsing under the pressure at great depths. The rib cage is flexible and yields to increased pressure without breaking. Finally, the spent carbon dioxide and nitrogen is returned to the lungs where it is kept at the same pressure as the sea outside. This prevents nitrogen bubbles from forming in the circulatory system and causing a condition known as decompression sickness, or "the bends".

## FEEDING

Whales are divided into two suborders, baleen whales and toothed whales, depending on their source of food. Baleen whales are equipped with distinctive sieves,

*A blue whale surfaces. Its spout can be 6–9 metres in height.*

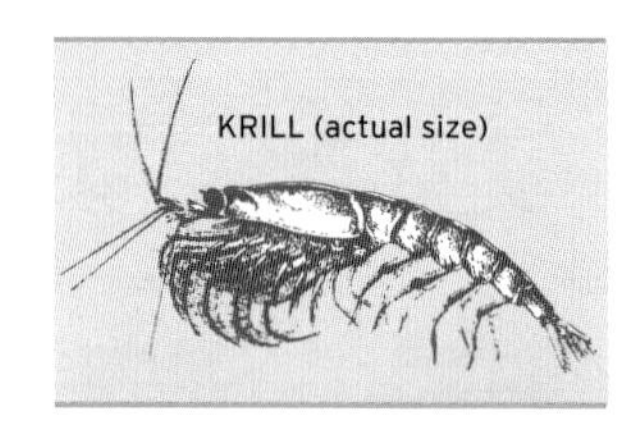


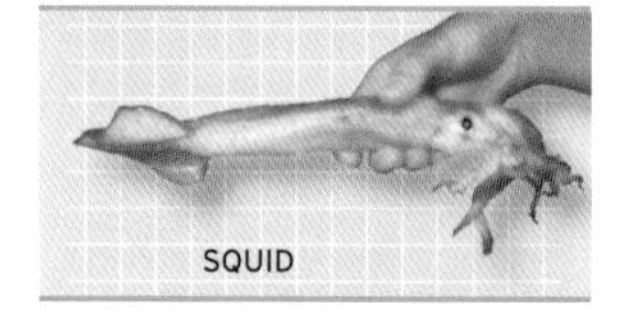

known as baleens, used to strain food from the sea. Baleens are ceratin plates growing from the roof of the whale's mouth. Each baleen is triangular in shape, with the longest side of the triangle frayed into a brush-like mass on the inside. The size of the baleen depends on the type of whale. They can be up to 400 in number and reach four metres in length, growing and renewing themselves throughout the whale's life. Baleen whales filter food from the sea by filling their mouth with water and then forcing it out again between the baleens with their tongue, trapping the food on the inside.

The main food of baleen whales consists of plankton, krill and various species of small pelagic fish. The baleen whales known as "rorqual" (blue whale, fin whale, sei whale, humpback whale and minke whale) have huge pleats extending from chin to belly. The pleats are reminiscent of a harmonica, enabling whales to gulp huge amounts of water. To give an example, a blue whale can accommodate 50 tons of sea in its mouth in order to consume the four tons of plankton it needs for subsistence every day over the summer months.

Toothed whales have single-rooted conical teeth. The number of teeth ranges from two to dozens. All the teeth are of the same kind and used to seize and tear the whale's prey, but not to chew. Most toothed whales live on fish, squid and octopi, although some species, such as orcas, also attack seals, sea lions, penguins and cormorants, and even other whales.

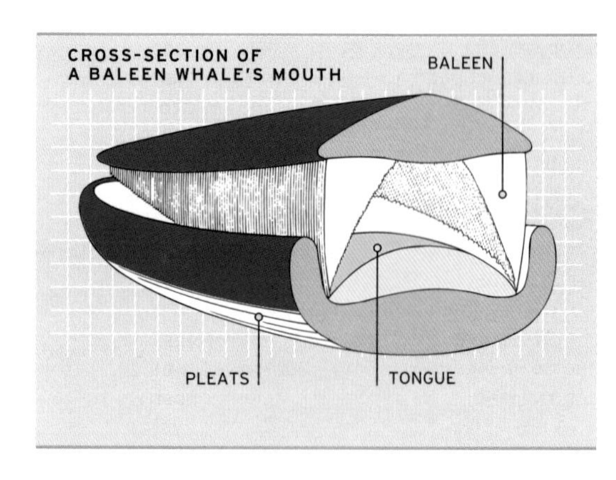


## Digestive System

Both baleen whales and toothed whales swallow their food without chewing, which complicates the digestive process. The stomach of the whale has adapted to this difficult task, and is divided into three parts.

Muscle tissue at the front of the stomach keeps the stomach in motion. It contains sand and broken shells (like birds' gizzards) which help the whale to crush its food. Actual digestion takes place in other parts of the stomach.

Research has shown that whales quench their thirst by drinking seawater. The salt is removed from the seawater by enormous kidneys.

Whales have extremely long intestines. In a twenty-metre sperm whale, the intestines can be up to 480 metres long. In some sperm whales a substance known as ambergris is formed in the digestive system. Ambergris is a hard lump originating in the indigestible remains of cuttlefish. It is formed in the intestinal tract and can cause digestion problems if the whale fails to excrete it. Ambergris is found in about one of every thousand sperm whales. For a long time it was in great demand for use in perfumes and sold at exorbitant prices.

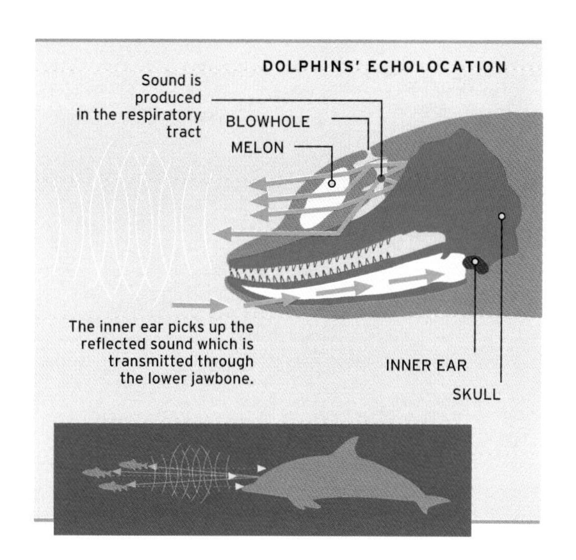


## NEURAL SYSTEM AND SENSORY ORGANS

The brain of the whale is more developed than the brains of most mammals. Dolphins are regarded as extremely intelligent, and they are quick to learn all kinds of tricks. Whales have no mucous membrane and therefore no sense of smell. The eyes are small, and it

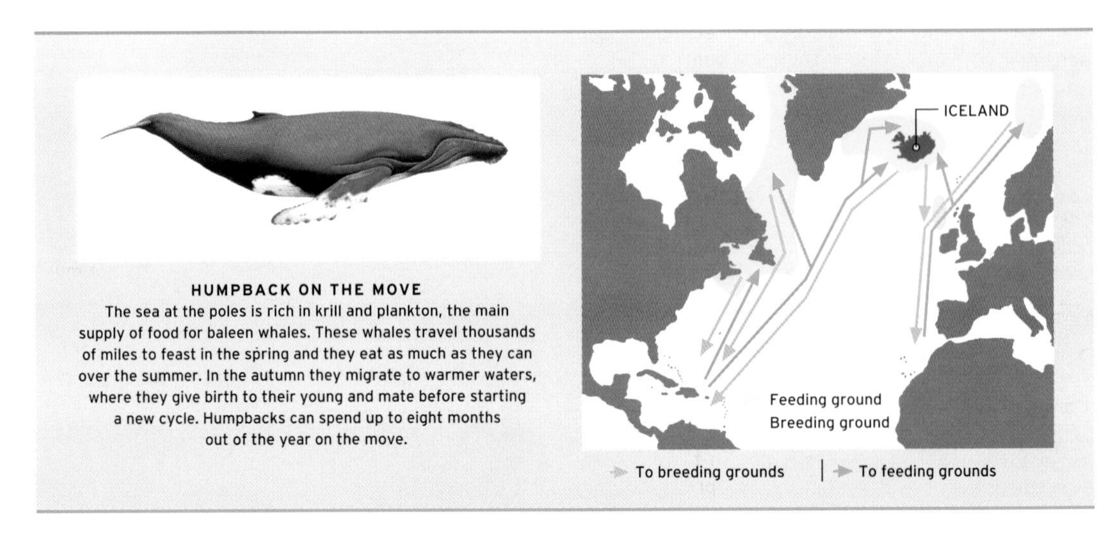


**HUMPBACK ON THE MOVE**
The sea at the poles is rich in krill and plankton, the main supply of food for baleen whales. These whales travel thousands of miles to feast in the spring and they eat as much as they can over the summer. In the autumn they migrate to warmer waters, where they give birth to their young and mate before starting a new cycle. Humpbacks can spend up to eight months out of the year on the move.

is not clear whether whales can distinguish between colours, but they have good eyesight in and out of the water.

### Echolocation

Although whales do not have visible ears, their hearing is their most important sensory organ. Whales emit and detect a wide range of high-frequency and low-frequency sounds. By analysing the echo of these sounds they get a picture of their surroundings, which enables them to orient themselves underwater, much like bats do in the dark.

Toothed whales emit a variety of sounds ranging to 280,000 Hz (cycles per second). Sounds emitted by baleen whales only reach 20,000 Hz, which makes them audible to humans. Whale sounds can reach a volume of eight decibels and carry hundreds of kilometres underwater. This applies particularly to the lowest frequencies. How whales manage to produce such a wide variety of sounds without vocal chords is an unsolved mystery.

**Communications**

The sounds are not only to help the animals orient themselves. Some scientists maintain that toothed whales can paralyse, and even kill their prey with high-frequency sounds. There appears to be little doubt that whales "converse" using sound; they can communicate with other animals of the same species at long distances and summon them or warn them of impending danger. The best known example of such communication is the song of the humpback whale, which is incredibly varied and beautiful, especially in the mating season. Some scientists have even maintained that they can distinguish between regional "dialects" of humpbacked whales. The song of the beluga whale is not as well known, but it is even more varied, and belugas have been called the "vocalists of the sea" or "sea canaries".

## REPRODUCTION

Since the genitals of whales are nearly always hidden in a fold of flesh on the whales' belly, they are not much help in sexing the animals. The difference in size between males and females is a better indication, as is the dorsal fin, e.g. of the orca. The mating season of most species of whales is in the autumn and the spring, but the actual time of mating is influenced by sea temperature and daylight time.

**Migrating Patterns of Whales**

Many species of whales have their regular habitat at high latitudes, but migrate long distances to warmer climates around the breeding season, e.g. the southern coast of Australia, or the coasts of Africa and South America. Some species spend months in warm seas where they raise their calves. In early summer, the

**DID YOU KNOW?**

Whales are always born fluke first, as otherwise they would drown.

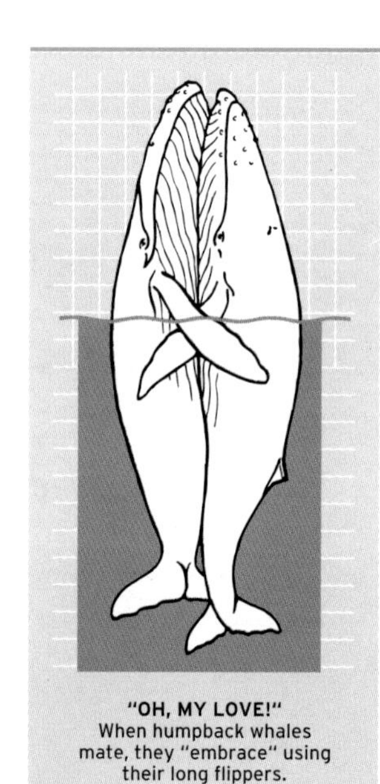

"OH, MY LOVE!"
When humpback whales mate, they "embrace" using their long flippers.

whales migrate north and south to Arctic and Antarctic waters to take advantage of the rich supply of food in those regions.

**Gestation and Calving**

The gestation period of whales varies, depending on species, from eight to eighteen months. Normally, female whales, referred to as cows, will calve every other year, bearing a single offspring. Twins are rare.

When the calf, which can weigh up to a ton, is born, the flukes always appear first. When the head and blow hole appear, the mother or another helpful whale will nudge the calf to the surface to help it take its first breath. Whales have no lips and cannot suckle. The cow therefore lies on its side and squirts the milk into the calf's mouth. Whale calves nurse from six to eighteen months, and the calves of the larger species will drink up to 250 litres of fat-rich milk (36% fat) each day. They grow rapidly and can gain up to 100 kilograms in weight in a single day. Toothed whales reach puberty about fifteen to twenty years of age, and most baleen whales at six years, except for the humpback, which attains puberty around the age of twenty.

Many species of whales can grow quite old, and baleen whales can reach the age of 60–70 years. The blue whale is believed to reach 90 to 100 years of age, and Greenland whales, or smooth-back whales, may grow even older.

*A pod of orcas. The male is distinquished by the tall fin, which can be up to 1.80 metres in height.*

## SOCIAL BEHAVIOUR

Social behaviour is an extremely prominent feature of some species of whales. Many toothed whales maintain family ties or form small groups known as pods, and even defend their hunting grounds against other pods. Family groups vary in size. In the case of orcas, pods are made up of females and their offspring. Male sperm whales are often accompanied by a "harem" of up to thirty cows, and there are reports of schools of pilot whales and dolphins with hundreds or even thousands of animals. Young bulls abandon their families, and the bachelors roam the oceans in groups. Baleen whales, on the other hand, are not gregarious animals, and their family patterns are unknown.

### Care

Whale cows care for their offspring from the time they begin to nurse and will risk their own lives in the defence of their young. The family plays an important role. Family assistance begins immediately at birth when older sisters or "aunts" assist in bringing the calf to the surface of the sea so that it can breath its first breath of life. Later, other members of the family will watch over the calf if the mother leaves it in search of food.

*Stranded sperm whales in Furufjörður fjord.*

### Stranding

Quite frequently, single whales or even entire schools of whales will swim or drift ashore. The reasons for this are not quite clear, but most scientists believe that the whales' sense of hearing could become distorted, e.g. as a result of parasites. If the leader of a school is afflicted, the entire school may follow and suffer the consequences.

**SLEEPLESS**

Whales cannot sleep.
If they do, they stop breathing.
A whale with its eyes closed
is probably half asleep.
When resting, whales
will swim slowly or languish
on the surface.

## Baleen Whales / Toothed Whales

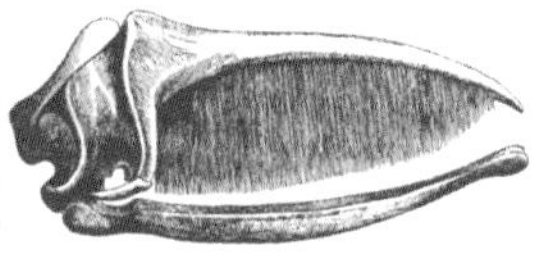

*Baleen whale skull.*

*Toothed whale skull.*

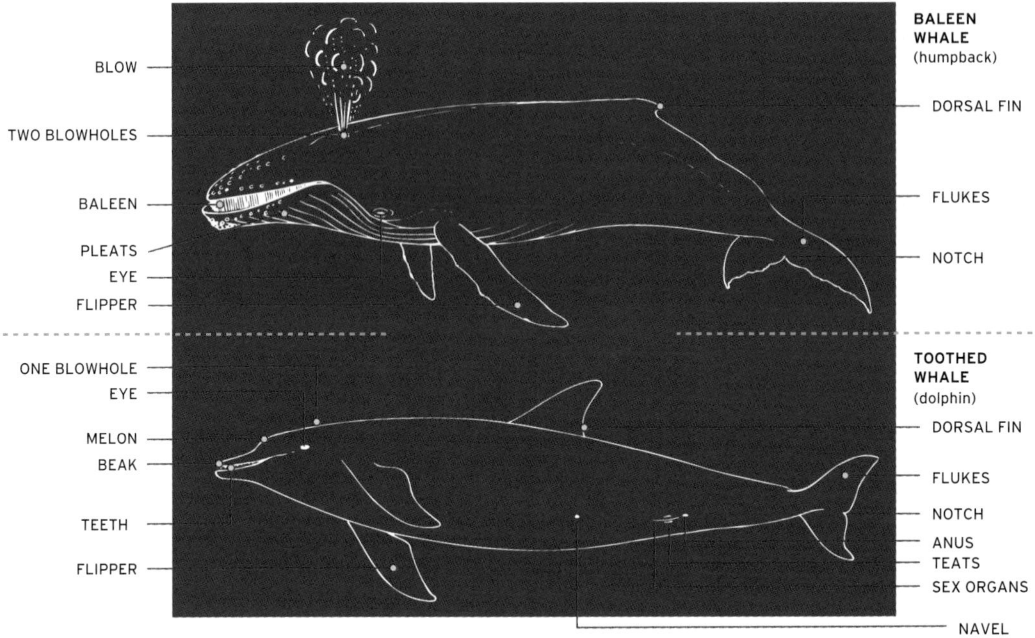

BLOW
TWO BLOWHOLES
BALEEN
PLEATS
EYE
FLIPPER

**BALEEN WHALE**
(humpback)

DORSAL FIN
FLUKES
NOTCH

ONE BLOWHOLE
EYE
MELON
BEAK
TEETH
FLIPPER

**TOOTHED WHALE**
(dolphin)

DORSAL FIN
FLUKES
NOTCH
ANUS
TEATS
SEX ORGANS
NAVEL

**PHOTOGRAPHING THE WHALES**
Best results are achieved using a camera with automatic focus and an 80–200 mm zoom lens, but any camera will do. 200 ASA film is suitable for most conditions.

*Minke whale poses for a portrait.*

# Blue whale

*Balaenoptera sibaldi*

| LENGTH | WEIGHT | ESTIMATED WORLD POPULATION | LIFE EXPECTANCY |
| --- | --- | --- | --- |
| 20–33 metres | 110–190 tons | 3.000–4.000 animals | 90 years |

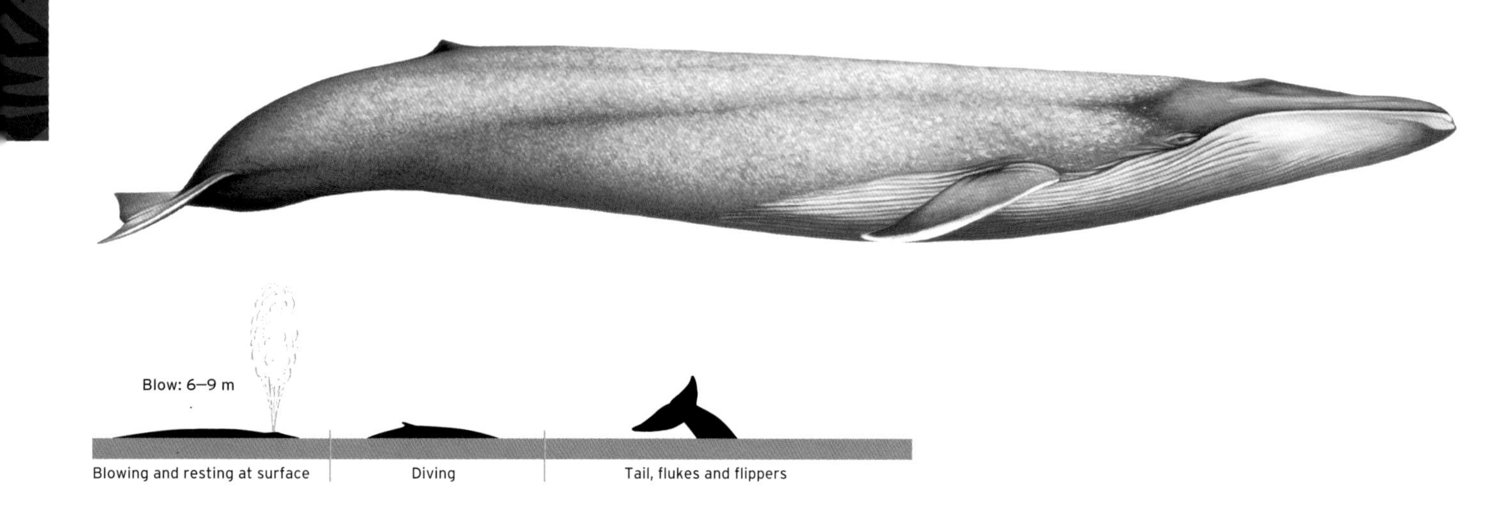

# Blue whale



The blue whale is the largest animal ever to live on Earth. The longest measured blue whale was 33.5 metres long, and the heaviest weighed almost 200 tons. Even though these giants are the largest animals on Earth, they survive almost entirely on some of the smallest living beings in the sea: plankton and krill.

One of the largest stocks of these majestic creatures, 700–1,000 animals, is believed to make its way to the coasts of Iceland in the summer. Blue whales approach Iceland in the spring and remain there into the autumn. Around breeding time, the blue whales migrate south, but their wintering grounds are unknown. They calve in the south seas, and the calves then accompany their mothers to Iceland in the spring. Their main feeding grounds around Iceland are off Breiðafjörður in western Iceland.

The spout is usually the first indication of the presence of whales. The spout of the blue whale is reminiscent of a geyser, and can reach heights of nine metres. Blue whales will surface three to five times in succession to breathe before they go into a deep dive which may last from seven to ten minutes, but frequently much more. Sometimes they will lift their flukes when diving, an action known as "fluking". The flukes can be up to 8 metres from end to end.

*Blue whales and the whale watching boat Haukur from Húsavík.*

**HEAVY HEART**

A blue whale weighs about 190 tons, which corresponds to about 27 elephants, 630 cows or 2,700 men and women.
A blue whale's heart weighs about 1,200 kilogrammes.
The tongue weighs about 7,000 kilogrammes.

# Fin whale *Balaenoptera physalus*



| LENGTH | WEIGHT | ESTIMATED WORLD POPULATION | LIFE EXPECTANCY |
|---|---|---|---|
| 20–25 metres | 50–80 tons | 120,000–150,000 animals | 90 years |

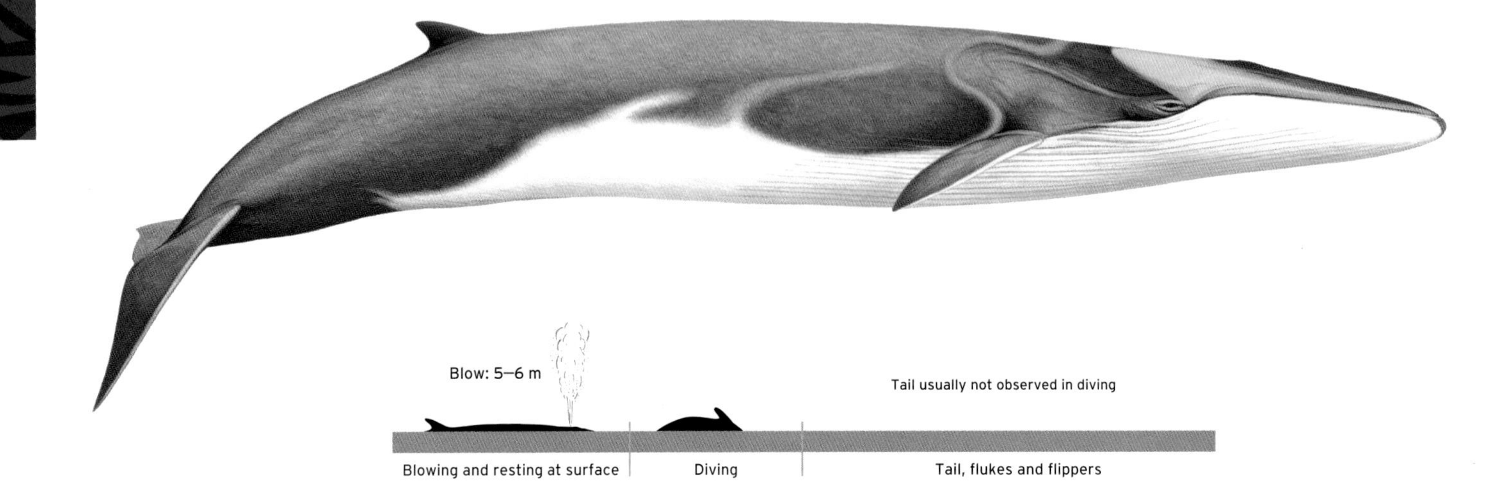

# Fin whale



Like the blue whale, the fin whale is a baleen whale; in fact the species are closely related.

Fin whales arrive in Iceland in the spring and remain until the autumn. It is believed that fin whales spend the winters in the distant ocean south and west of Iceland. The number of fin whales in Icelandic waters has been estimated at approximately 16,000–19,000 animals.

Fin whales can be identified by their colour. The right jaw is a distinctive white or pale colour, while the left jaw is dark grey or black. The same is true of the baleens. The whale itself is dark, but with lighter, brown-toned streaks or swirls across the back, reminiscent of "northern lights".

The spout is powerful and can be seen at long distances when conditions are right. The fin whale surfaces to breathe three to five times before sounding. Fin whales usually remain submerged for five to eight minutes at a time, but frequently longer. Fin whales rarely fluke when they dive. When feeding on plankton and krill they often roll over at the surface.

*Fin whale surfaces during a whale watching tour.*

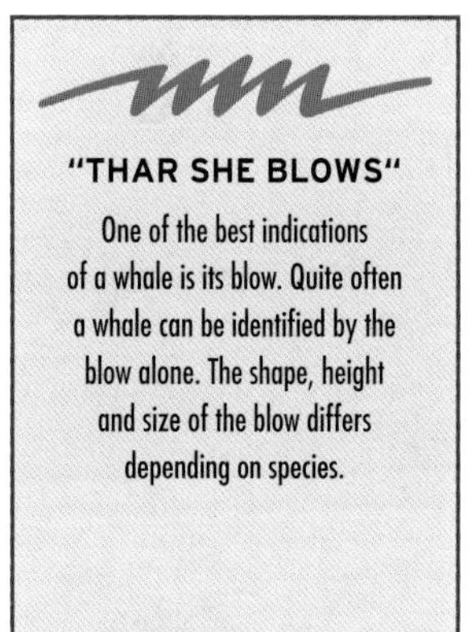

# Sei whale

*Balaenoptera borealis*

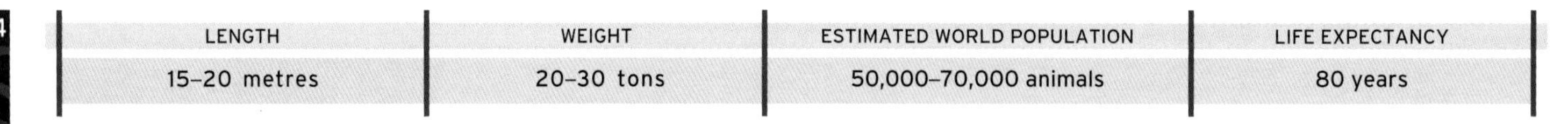

| LENGTH | WEIGHT | ESTIMATED WORLD POPULATION | LIFE EXPECTANCY |
| --- | --- | --- | --- |
| 15–20 metres | 20–30 tons | 50,000–70,000 animals | 80 years |

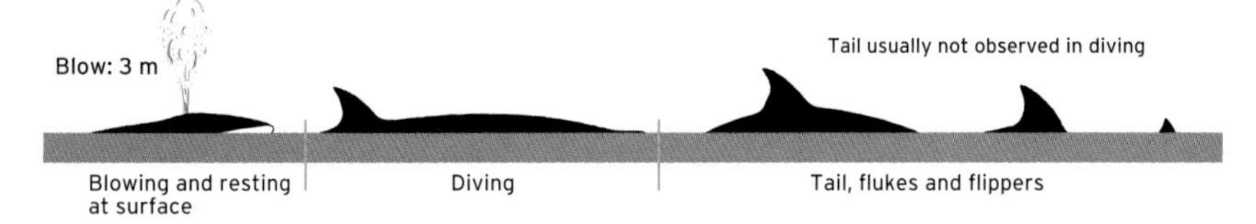

## Sei whale



The sei whale is a baleen whale, like the blue whale and fin whale.

Sei whales arrive in Icelandic waters in the spring, like most other large whales, after wintering in the South Seas. They normally keep to deeper waters than, e.g., fin whales, about 30–60 miles west of Iceland. Although sei whales mainly remain in this area, they have also been spotted off the north and south coasts of Iceland. An estimated 10,000 sei whales travel to Icelandic waters in the summer. Sei whales feed mainly on plankton, krill and various species of small schooling fish.

The sei whale is very dark in colour, like the minke whale.

Their spout is easily spotted and can be observed for quite a while in good visibility. A sei whale can easily be mistaken for a large minke whale. In cases of doubt, the sei whale can be identified by its size and the unusually tall dorsal fin. Sei whales come up for air three to four times in succession. They normally stay underwater for five to seven minutes at a time, but sometimes for much longer. Sei whales do not fluke when they dive, but will often roll over on the surface when feeding.

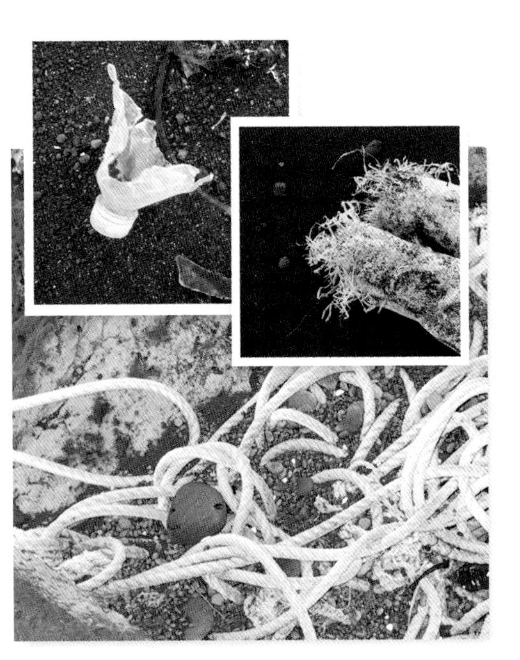

### POLLUTION

Every single day, an enormous quantity of waste and polluting materials are thrown into the sea. Tossing things into the sea puts them out of sight – not out of existence. **"Think before you throw"** is the order of the day.

# Humpback whale *Megaptera novaeangliae*

| LENGTH | WEIGHT | ESTIMATED WORLD POPULATION | LIFE EXPECTANCY |
| --- | --- | --- | --- |
| 13–17 metres | 25–40 tons | 10,000–15,000 animals | about 95 years |

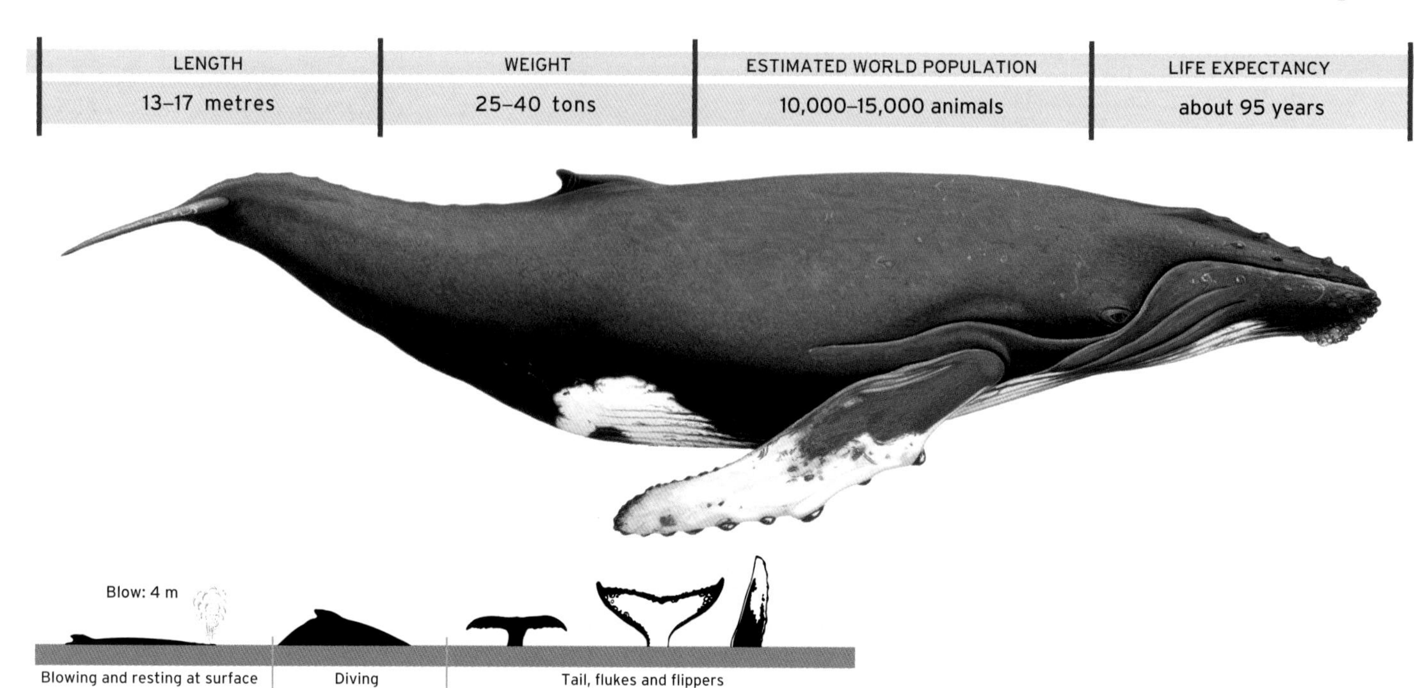

# Humpback whale



Humpback whales arrive in Icelandic waters in spring, like most large whales. They travel from the southern oceans where they winter, e.g. in the Bay of Mexico. Humpback whales mainly frequent shallow waters around Iceland and often come into fjords and bays in search of food. Their number is uncertain, but it is estimated that 1,500 to 1,800 animals enter Icelandic waters each summer. Humpback whales feed primarily on plankton, krill and small fish, such as capelin.

The enormously long flippers of the humpback whale are a major identifying feature of the species. These can be up to five or six metres in length, and the whales occasionally raise them out of the water and slap the surface, apparently in play. They sometimes also roll over on the surface and even leap out of the water with a great deal of flipper-slapping. The head and flippers are covered in barnacles, which attach themselves to the whale shortly after birth and remain there during the whale's lifetime.

Humpback whales have a powerful spout that lasts for quite a while. They surface to breathe three or four times before sounding. Normally, they remain underwater for five to seven minutes at a time, but often for much longer. When humpback whales take sounding dives they nearly always raise their flukes out of the water, revealing their unique "fingerprints", a distinctive white pattern at the bottom of the tail fin that can be used for the identification of individuals.

*Humpback whale in the midnight sun.*

Photographs of humpback flukes are therefore important data. We encourage readers to preserve any photographs they may have of humpback flukes, along with data on where and when they were taken. The Húsavík Whale Centre collects data and photographs of whales and is grateful for all donations of photographs.

*Humpback whales have the longest flippers of all whales. They can reach up to 5–6 metres in length.*

# Minke whale

*Balaenoptera acutorostrata*



| LENGTH | WEIGHT | ESTIMATED WORLD POPULATION | LIFE EXPECTANCY |
|---|---|---|---|
| 7–10 metres | 8–10 tons | uncertain, at least several hundred thousand animals | about 50 years |

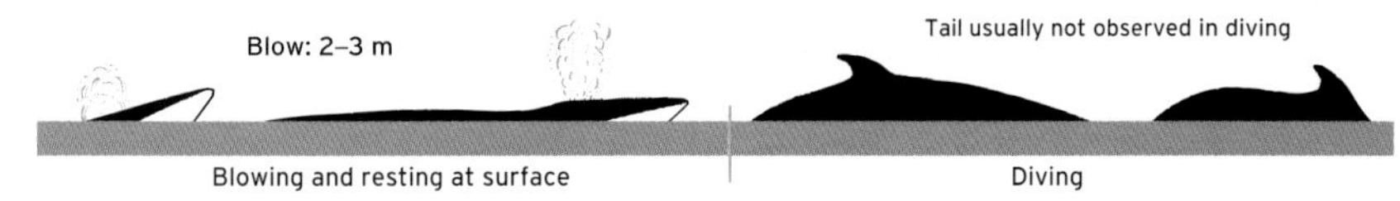

# Minke whale



Despite its considerable size, the minke whale is the smallest of the baleen whales. It may reach up to 10 metres in length and weigh 10 tons. Males are usually smaller, about seven or eight metres in length. Minke whales in northern seas have white "bracelets", diagonal stripes across their flippers, unlike whales of the same species in southern seas which have all-black flippers.

The Icelandic minke whale population is an estimated 50,000–60,000 animals, most of them resident in Icelandic waters. Minke whales feed mainly on plankton, krill and small surface fish, but they are believed to base part of their diet, about 1–6%, on various stockfish, such as cod.

The odds of spotting minke whales on whale watching excursions in Icelandic waters are excellent, about 95–99%. Minke whales in these waters are extremely inquisitive, playful and entertaining. They often come up to the whale watching boats to observe their observers or examine the boats, which they have come to know after an acquaintance going back many years. When minke whales are feeding they often exhibit nothing but a brief spout, a black back and high dorsal fin before submerging.

Minke whales spout and breathe three or four times in succession before diving. They can remain under water for up to 20 minutes at a time, but normally they stay under for three to five minutes. They do not fluke, but often hump their backs before vanishing into the deep. Minke whales sometimes leap out of the water and then often many times in succession.

*Minke whale observing whale watchers.*

# Sperm whale *Physeter macrocephalus*

| LENGTH | WEIGHT | ESTIMATED WORLD POPULATION | LIFE EXPECTANCY |
|---|---|---|---|
| 12–18 metres | 30–50 tons | uncertain, at least several hundred thousand animals | about 70 years |

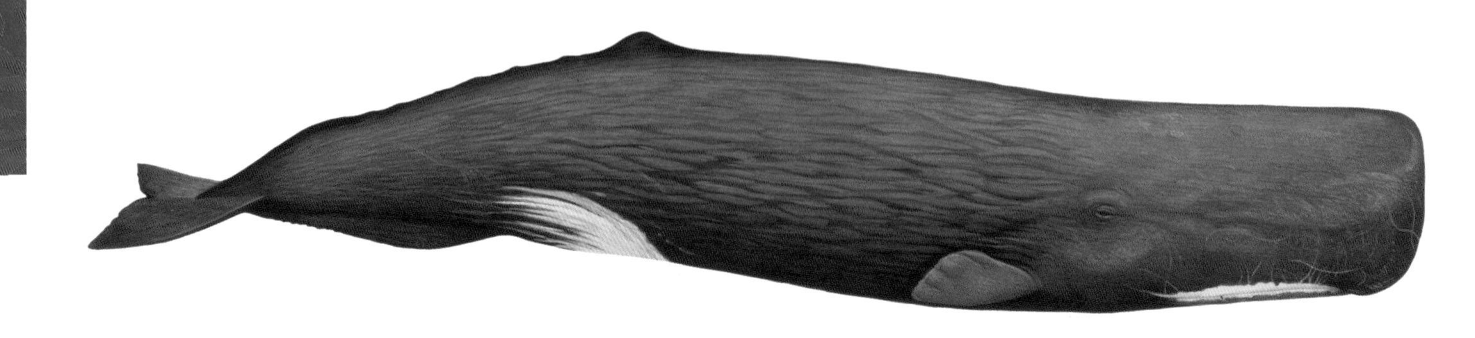

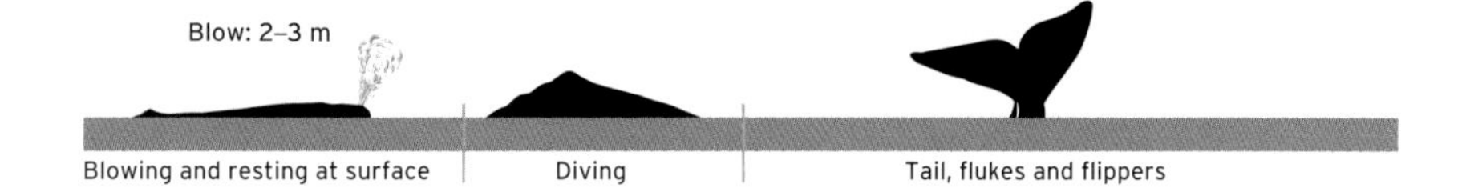

# Sperm whale



Sperm whales are the largest of the toothed whales. They arrive in Icelandic waters in the spring after wintering in the South Seas. Sperm whales stay mostly in deep waters, where feeding conditions are good. The 1,200–1,400 animals that come to Iceland in the summertime are single bulls that have lost in the competition for females. The females remain in the South Seas with the strongest males.

The most prominent feature of sperm whales is their enormous head, which can be up to one third of the whale's overall length. The spout of sperm whales is quite tall and powerful, and slants to the left from the blowhole, which is located on the left side of the head.

Sperm whales can remain underwater for a long time, more than an hour. However, the sounding dives commonly last for 30–40 minutes. When the whales come up for air they often remain half-submerged on the surface for five to ten minutes before sounding again. They normally raise their flukes out of the water as they disappear into the deep.

The diet of sperm whales consists chiefly of squid, redfish, black halibut, monkfish and other species of fish, but in the South Seas, octopus frequently features on their menu.

*A sperm whale in its death throes. Illustration from a book by Thomas Beale dating from 1821.*

# Northern bottlenose whale *Hyperoodon ampullatus*



| LENGTH | WEIGHT | ESTIMATED WORLD POPULATION | LIFE EXPECTANCY |
|---|---|---|---|
| 7–9 metres | 6–8 tons | unknown | about 50 years |

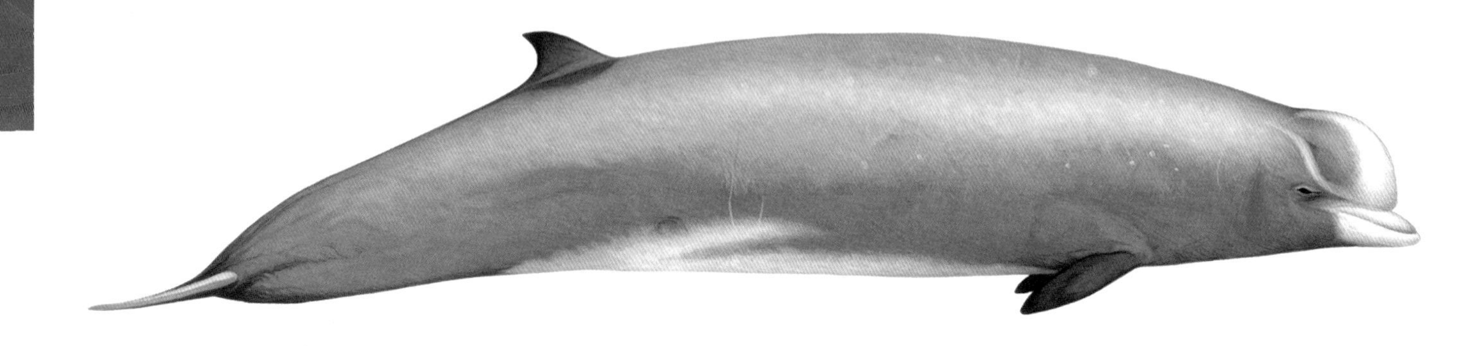

# Northern bottlenose whale



The northern bottlenose whale is about the same size as the minke whale, but in other respects the two species are very different. Northern bottle-nosed whales are brownish in colour and are rarely seen in inlets and bays, preferring deep waters. No other whale species, except sperm whales, sound as deep as the northern bottlenose whale. Icelandic scientists believe the northern bottlenose whale population in Icelandic waters is about 40,000 animals. They mostly keep to deep waters south-east of Iceland in winter, venturing closer to the shore in summer. The diet of northern bottlenosed whales, like that of other toothed whales, consists mostly of fish.

Northern bottlenose whales can remain submerged for more than an hour, but they normally surface within half an hour of diving. Their spout is low and difficult to spot. The whales spout three or four times in succession before sounding. When diving they often hump their back before taking the plunge. Northern bottlenose whales rarely fluke, but have been known to leap.

*Rare sight. Northern bottlenose whale leaping in Skjálfandaflói bay.*

**BUNDLE UP!**

It can be cold at sea, and you are quite likely to get wet. As far as warm clothes are concerned too much is better than too little. There is always space aboard for extra clothing. In good weather, don't forget your sunscreen to protect yourself from the sun reflected by the sea. It is a good idea to bring along a plastic bag for your camera.

# Long-finned pilot whale *Globicephala melas*

| LENGTH | WEIGHT | ESTIMATED WORLD POPULATION | LIFE EXPECTANCY |
|---|---|---|---|
| 4–8 metres | 2–5 tons | uncertain, at least several hundred thousand animals | about 40 years |

# Long-finned pilot whale



Long-finned pilot whales are members of the dolphin family and common in Icelandic waters. Their number around Iceland is estimated at about 35,000 animals. Long-finned pilot whales often travel in large groups, or "schools", although individuals are sometimes seen travelling alone. They are most frequently sighted off the south-east, south and west coasts of Iceland in late summer and autumn.

Long-finned pilot whales have a tall dorsal fin on the front half of their back. They have small flukes which are usually visible when the whales dive.

Long-finned pilot whales usually stay underwater for five to ten minutes at a time, but can remain submerged for longer periods. They are known to have gone as deep as 600 metres, but normally they search for food – mainly squid and various species of fish – closer to the surface, at depths of about 30–60 metres. On whale watching excursions large schools of long-finned pilot whales are sometimes seen swimming slowly near the surface, but they are not renowned for their aerobatic skills.

*In October 1986 these pilot whales were stranded on a beach by Þorlákshöfn, on the south coast of Iceland.*

# Orca (killer whale)

*Orcinus orca*

| LENGTH | WEIGHT |
| --- | --- |
| 6–9 metres | 3–9 tons |

| LIFE EXPECTANCY |
| --- |
| male about 50 years<br>female about 60–70 years |

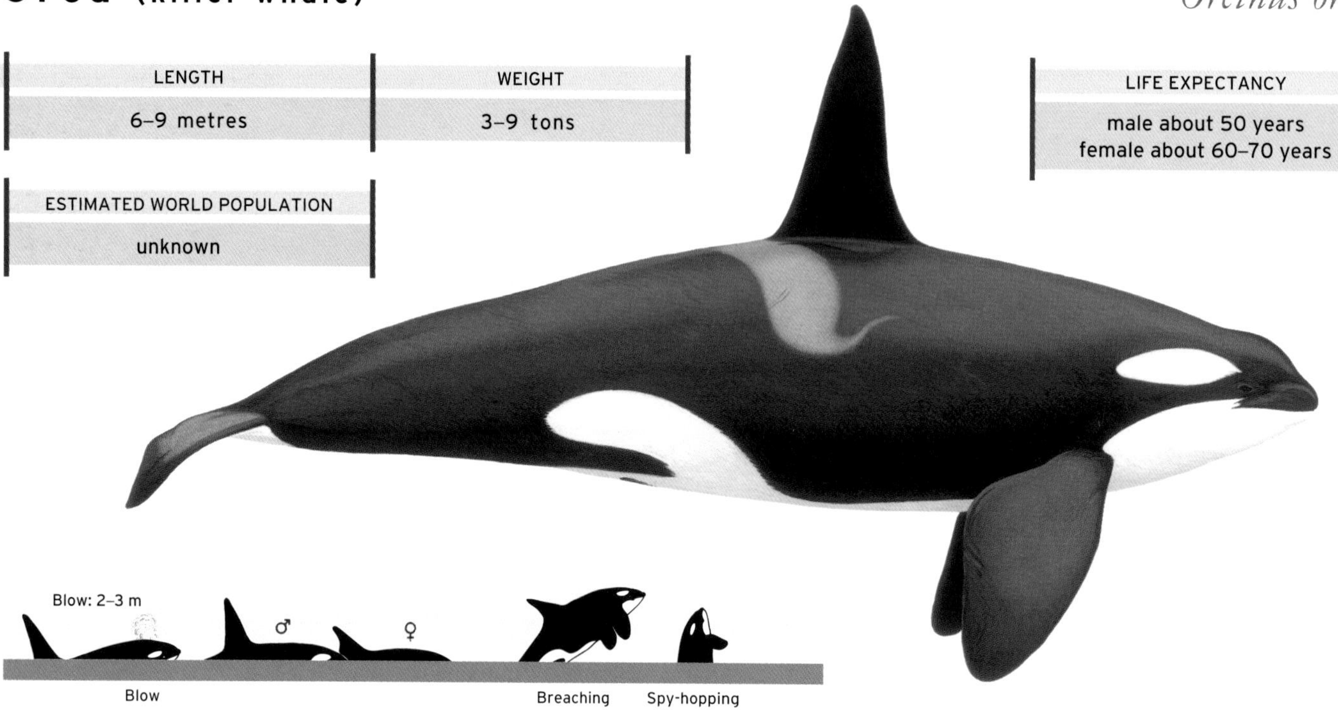

| ESTIMATED WORLD POPULATION |
| --- |
| unknown |

Blow: 2–3 m

♂ ♀

Blow

Breaching

Spy-hopping

## Orca (killer whale)



The orca (killer whale) is probably one of the best-known whale species in the world, particularly as a result of the attention generated by captured killer whales in oceanaria across the world. The most famous orca of all time, Keiko, the star of the *Free Willy* movies, returned to his Icelandic home waters in the autumn of 1998 when he was brought to the Westman Islands on a U.S. Air Force cargo plane. Keiko has made excellent progress since his arrival, and is now under the constant care of the staff of Ocean Futures, (there is more on Keiko at www.oceanfutures.org).

Orcas are members of the dolphin family. They are "family animals" and stick together in small groups, known as pods. These family groups are made up of females and their offspring. There are no fathers as such in the pods, as the males leave the pod over the mating season to mate with females from other groups and then return to their own pod.

The Icelandic orca population is believed to be resident and to number about 5,000 animals. Orca pods can be spotted all around Iceland, most frequently in the herring grounds off the East Fjords and the South and West of Iceland. In summer, orcas are often seen closer to land, inside inlets and bays, but they prefer deeper waters in winter. The whales are easily identified by their black and white colouring and the tall dorsal fin, which in males can reach a height of 1.8 metres.

Orcas are the fastest of all whales and can swim at a speed of about 50 km per hour. They can remain submerged for up to 20 minutes at a time, but normally surface in about five minutes. They spout three or four times in succession before sounding.

Orcas in Icelandic waters subsist mostly on herring, but they have been known to attack seals and even other whales.

*Keiko, the world's most famous orca.*

# White-beaked dolphin

*Lagenorhynchus albirostris*

| LENGTH | WEIGHT | ESTIMATED WORLD POPULATION | LIFE EXPECTANCY |
|---|---|---|---|
| 2.5–3 metres | 180–350 kilos | unknown | about 25 years |

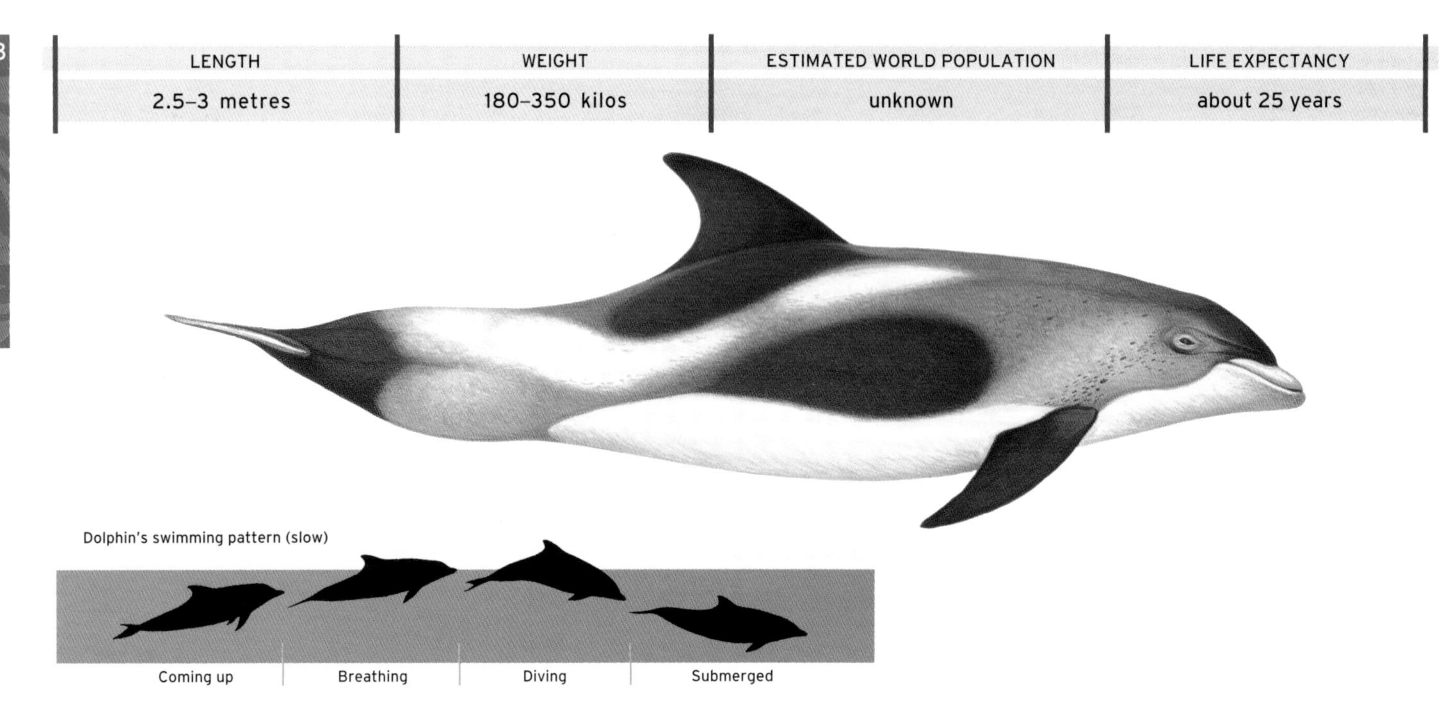

# White-beaked dolphin



The white-beaked dolphin is the single most frequently sighted dolphin on whale watching trips. The population in Icelandic waters is estimated at 10,000 to 12,000 animals. It is believed to be resident in Icelandic waters. Large groups of white-beaked dolphins are occasionally seen in fjords and bays, but usually they travel in smaller groups of five to ten animals.

Like other dolphins, the white-beaked variety does not normally remain underwater for long. They stick to the surface, and are constantly on the move. Dolphins feed primarily on various small fish, such as mackerel and herring, and squid.

White-beaked dolphins are extremely fast swimmers and often come leaping towards ships and boats to frolic in their bow-waves. Watching them use their tail stocks and flukes to propel themselves out of the waves to spout is a fascinating sight. They often leap straight out of the water, as if in play, and then continue to leap incessantly until they tire and wander off.

*Dolphins sometimes get caught in nets and die. This one has clearly had his lower jaw entangled for quite some time.*

**TEETH**

Unlike our teeth, which come in various sizes and shapes, dolphins' teeth are all the same shape.

# Atlantic white-sided dolphin *Lagenorhynchus acutus*

| LENGTH | WEIGHT | ESTIMATED WORLD POPULATION | LIFE EXPECTANCY |
|---|---|---|---|
| 2–2.5 metres | 130–230 kilos | unknown | about 25 years |

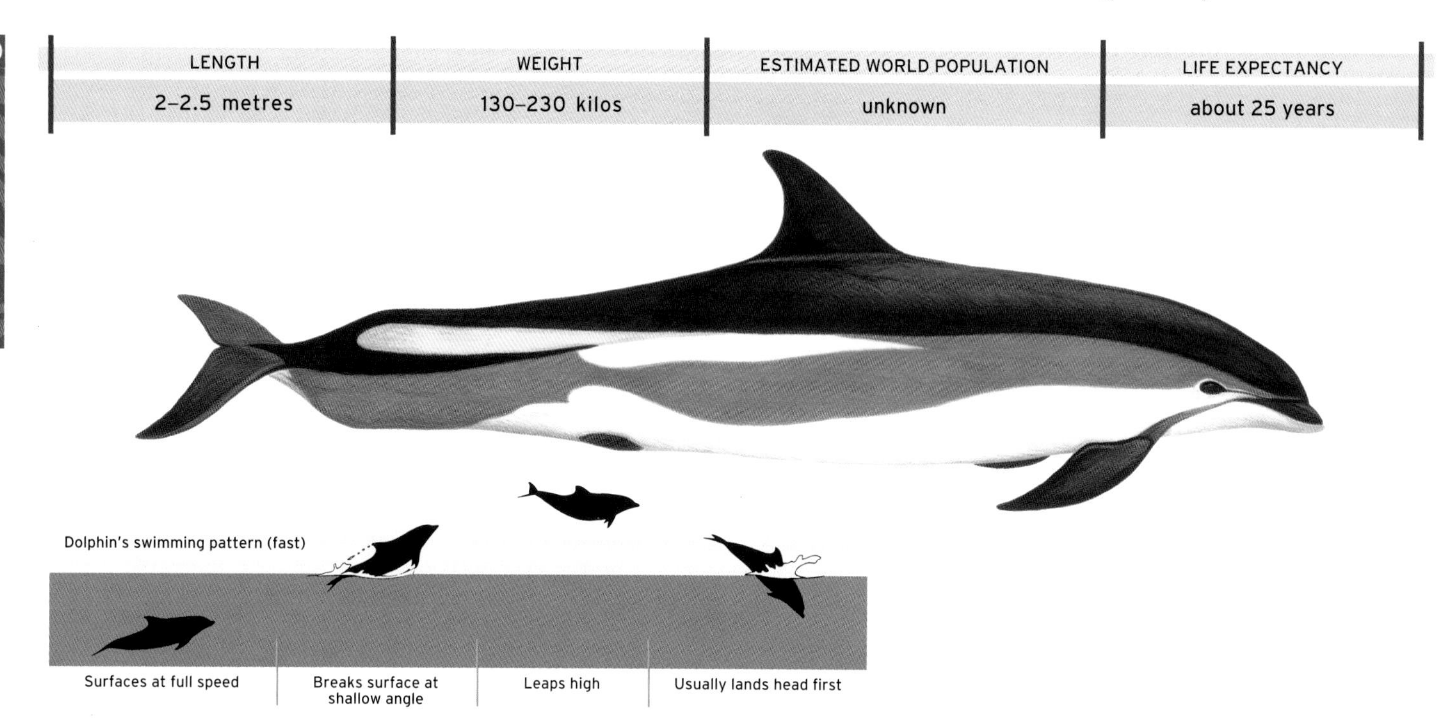

# Atlantic white-sided dolphin



White-sided dolphins are quite common in Icelandic waters. They often travel together in large groups, although sometimes they are seen travelling alone. The most frequent sightings are off the south, south-west, west and north-west coasts. They are closely related to the white-beaked dolphin, but somewhat smaller. The population in Icelandic waters has been estimated at about 35,000 to 38,000 animals. Their food, like that of other dolphins, consists primarily of various small fish, such as mackerel and herring, and squid.

The white-sided dolphin is most easily distinguished from its closest relatives by the colouring. Like most dolphins, white-sided dolphins are quite dark – almost black on the back and flippers, and grey on the side. Their principal identifying features are the yellow and white stripes and spots on their sides. These can be seen from close up.

Like other dolphins, white-sided dolphins do not remain submerged for long. They stick to the surface and are constantly on the move. The white-sided dolphin is an extremely fast swimmer, and is often seen leaping straight out of the water, as if in play, usually landing on its side.

*Mark Carwardine cought this photograph of a frolicking dolphin during a whale watching expedition.*

**SAY CHEESE!**

The shape of the dolphin's mouth makes it look like it is always smiling.

# Harbour porpoise

*Phocoena phocoena*

| LENGTH | WEIGHT | ESTIMATED WORLD POPULATION | LIFE EXPECTANCY |
|---|---|---|---|
| 1.5–2 metres | 55–70 kilos | unknown | about 30 years |

# Harbour porpoise



The porpoise is the smallest of all whale species in Icelandic waters and also very common. The local population is probably about 25,000–27,000 animals. They often form large groups, although sometimes they are also spotted alone. Porpoises prefer shallow waters and are therefore often seen in fjords and bays. There have also been sightings in estuaries.

Porpoises are toothed whales, like dolphins, but belong to a separate genus. Their diet consists primarily of various small schooling fish, but also herring and capelin. They sometimes try to snatch fish from nets, often paying the price with their lives, as they frequently get tangled in nets and drown.

These small whales can be difficult to spot. There is often a sudden glimpse of a small dorsal fin, but then the porpoise swims off at high speed, disappearing in the wink of an eye.

On whale watching trips, groups of porpoises are frequently seen bustling through the water. They tend to avoid ships and boats, although they have been known to approach whale watching boats with their offspring.

*Porpoises quite commonly get entangled in fishing gear. The photograph shows a porpoise and a seal who got entangled in a net and drowned.*

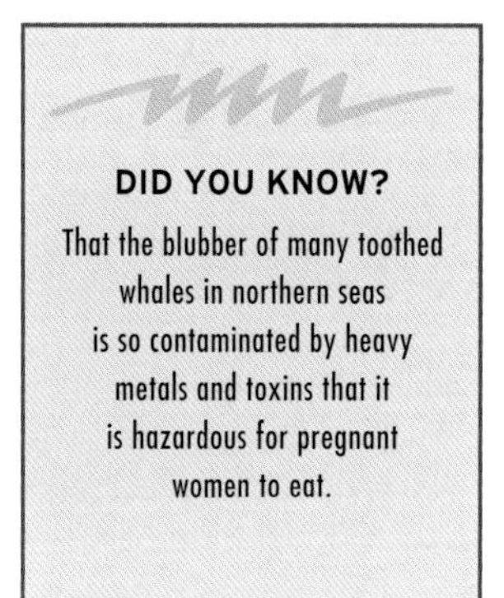

# Whaling in Icelandic Waters

*Stranded whale. Illumination of the mediaeval manuscript* Jónsbók *(14th. century).*

Ever since the settlement of Iceland, whales and their utilisation have been vital to the survival of Icelanders. The word hvalreki, literally "beached whale", now means a "lucky find" or lottery win, but in olden times it had a more literal meaning. In meagre years, finding a beached whale could mean the difference between life and death by starvation for many people. When a whale was found, the finder was under obligation to report the find to his neighbours. The whale was then divided according to fixed rules. There are records from ancient times of disputes, and even deadly feuds, over beached whales.

Basques from the north of Spain and the south of France were the first Europeans to engage in organised whaling. As early as the 12th century they began to hunt right whales as they skirted the Bay of Biscay in large schools on their route from the Arctic Ocean. These slow swimmers were easy to harpoon and their blubber was melted for use as lamp fuel and to make candles and soap. Later, after these whales had disappeared from the coasts of Spain as a result of excessive hunting, the Basques began to search for them, sailing all the way to Iceland and even Newfoundland.

There are sources from the 16th and 17th centuries testifying to the extensive whaling conducted by the Basques in Icelandic waters. Icelanders would trade with the Basques, but their involvement in the hunting was limited as they were poorly equipped to harpoon whales, in a process known as "ironing". When a whale was harpooned, there was no way to tow it ashore, and the hunters simply had to hope the dead whale would wash up on the beach. Harpoons usually bore the owner's registered initials so that a dead whale could be traced to the man who harpooned it – who was entitled to the lion's share of the whale if it was found.

Whales and dolphins trapped in fjords by drifting ice from the Arctic were usually killed with harpoons or other blades. This was usually a communal affair.

*Whaling in northern seas. Detail from an anonymous German painting dated 1776.*

The carcasses were then butchered and the meat divided among the local farms. It was also a common practice to drive schools of small whales ashore for their meat.

In the 17th and 18th centuries, the Dutch, English, Danes, Germans and Norwegians began to engage in whaling on a large scale. In addition to the oil, the baleen plates became coveted merchandise. Man has always been a slave to fashion, and this time it was the baleen that was used to make corsets for high-society women.

The Dutch conducted most of their whaling in the Arctic Sea. Each summer they operated a Svalbard whaling station known as Smeerenburg. Thousands of people worked there. These whalers hunted the bowhead, which abounded in the area. When the population had been depleted, the whalers moved west to Jan Mayen and Greenland waters.

The appearance of steamships in the mid-19th century enabled whalers to go after faster swimming whales than before, e.g. rorquals and sperm whales. The problem, however, was that these types of whales sink when dead, unlike the slower whales, which have a much thicker layer of blubber and therefore tend to float. The Norwegian Svend Foyn soon developed a new method of hunting. The new method was to harpoon the whales using a gun, and after the whale had been

*Whaling gun with explosive harpoon.*

killed by an explosive in the point of the harpoon, air was pumped into the carcass to keep it afloat. This method enabled organised hunting of the largest species of whale, the blue whale and fin whale. The carcasses of whales killed in this way were towed ashore and processed in large whaling stations.

In 1883, Norway was granted permission by Icelandic authorities to set up whaling operations in Iceland. At first, eight stations were set up in the West Fjords, but five were added later on the east coast of Iceland. To give an idea of the extent of the whaling, a total of 1,305 whales were processed in these stations in 1902. Two whaling stations belonging to the Norwegian Hans Ellefsen, in western and eastern Iceland, were the largest whaling stations in the North Atlantic in those years. According to sources dating from about 1900, one sperm whale yielded about ten tons of fat, three tons of meal processed from the whale meat and seven tons of bone meal. The remainder of the carcass was simply dumped on the shore. The resulting pollution caused the locals to grumble about the activities of the Norwegian whalers, as it was a hazard to sheep that wandered down to the shore in quest of seaweed.

Whaling off the East Fjords only lasted slightly more than a decade. Shortly after 1913 there were not enough whales left to sustain the whaling operations. In the end, the Norwegian whalers moved their activities to the Antarctic, where there was plenty of whales and the operations more lucrative.

By 1915, the Norwegians had killed about 17,000 whales. It was clear that this had cut deeply into the stocks of large whales in Icelandic waters, and in 1915 the Icelandic parliament passed a law on the protection of whales in Icelandic waters. This was the first time in history such a law was imposed. The law was repealed in 1928 when it was believed that the whale population had reached sustainable numbers again after the relentless slaughter of the years before 1915.

**Icelandic Whaling**

Icelanders began whaling in the modern sense in 1935 on the basis of new legislation on whaling in Icelandic waters. The new legislation granted to Icelanders the exclusive right to hunt whales in Icelandic territorial waters, and furthermore decreed that all whales killed should be fully utilised. The new law provided the foundation for the first Icelandic whaling station, which was established in Tálknafjörður and operated in the years 1935–1939. The whaling station in Hvalfjörður (Whale Fjord) was started in 1948. During the four decades that the station was operated, about 300–400 whales were processed every year, i.e. a total of some 15,000 whales.

In 1950–1985, Icelandic whaling was mostly limited to hunting fin whales, sei whales and minke whales. There was some hunting for blue whales, sperm whales and humpbacks for the first few years, but this was stopped by an act of law when it was discovered that stocks were being depleted so rapidly that there was a danger of their disappearing altogether from Icelandic waters. Icelanders had themselves witnessed the extermination of the northern right whale at the hands of Norwegian and other foreign whalers around the year 1900 and did not want to see this happen to other whale species. The northern right whale is not the only species of whale to disappear from Icelandic waters. It is believed that the grey whale was once common around Iceland, but now the species has totally disappeared from the Atlantic Ocean and is found only in the Pacific. The grey whale, which feeds mostly on molluscs and crabs in shallow waters, was no doubt an easy prey for the whalers in early centuries.

*Whale butchered at the Hvalfjörður Whaling Station. Icelanders stopped whaling in 1989.*

**Whaling Ban**

In 1983 the International Whaling Commission passed a ban on whaling with effect from 1986. Iceland did not protest the ban and was allowed to catch 60 large whales each year for scientific purposes. The scientific whaling ended in 1989. Numerous foreign environmental organisations had already harshly criticised this hunting.

*Landing whales in Hvalfjörður.*

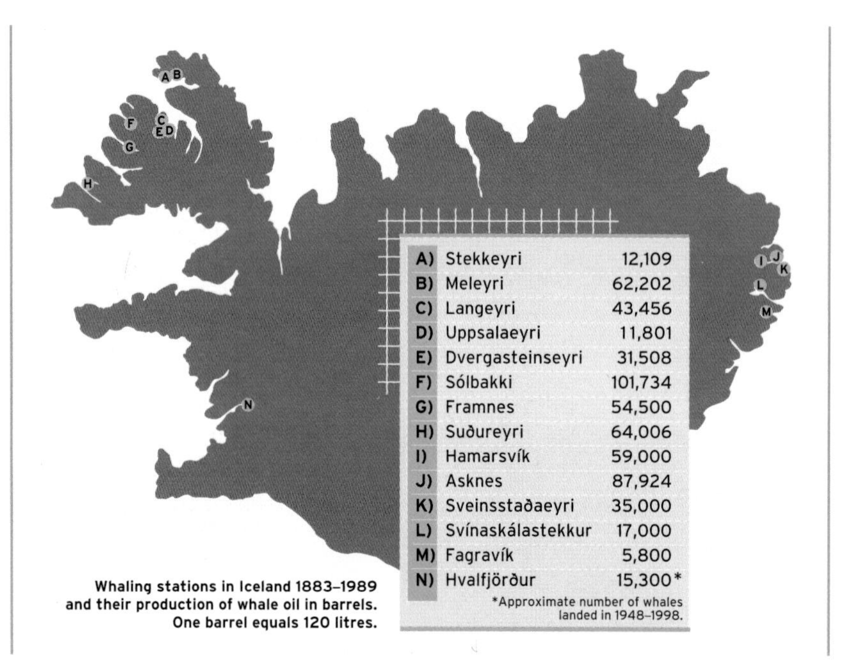


| |  | |
|---|---|---|
| A) | Stekkeyri | 12,109 |
| B) | Meleyri | 62,202 |
| C) | Langeyri | 43,456 |
| D) | Uppsalaeyri | 11,801 |
| E) | Dvergasteinseyri | 31,508 |
| F) | Sólbakki | 101,734 |
| G) | Framnes | 54,500 |
| H) | Suðureyri | 64,006 |
| I) | Hamarsvík | 59,000 |
| J) | Asknes | 87,924 |
| K) | Sveinsstaðaeyri | 35,000 |
| L) | Svínaskálastekkur | 17,000 |
| M) | Fagravík | 5,800 |
| N) | Hvalfjörður | 15,300* |

*Approximate number of whales landed in 1948–1998.

**Whaling stations in Iceland 1883–1989 and their production of whale oil in barrels. One barrel equals 120 litres.**

Many environmentalists regarded hunting for scientific purposes as a mere façade, designed to circumvent the whaling ban, as all products were sold directly to the Japanese market. Minke whale hunting, which had gone on sporadically since 1914, ceased in 1984.

When the International Whaling Commission rejected the opinion of the Scientific Committee that whaling should be permitted again, Iceland resigned from the IWC in 1991. Unlike Iceland, Norway and Japan had protested the whaling ban as early as 1985 and were therefore not bound by it and could continue to hunt whales for scientific purposes.

As long as Iceland remains outside the IWC it is hard for Icelanders to resume whaling again. Despite pressure from various people and some members of Parliament, no whaling has been allowed since 1989 and it is difficult to see how and when Iceland can start whaling again. In order for that to happen the country needs to rejoin the IWC and the ban on sales of whale products must be lifted. The effect whaling would have on whale watching must also be taken into consideration as it is a fast growing industry in Iceland. The best use of whales as a natural resource might therefore be to keep them where they are – in the sea.

# Research and Whale Watching



In the spring of 1995, a three-day course was given in Keflavík by WDCS, a British environmental organisation. English whale watching experts Mark Carwardine, Alison Smith and Erich Hoyt introduced several ideas that paved the way for whale watching in Iceland. They also gave a course for Icelanders interested in offering whale watching tours as a tourist attraction. The purpose of the course was to ensure sensible long-term arrangements for this new aspect of Icelandic tourism, which has flourished in recent years.

From the beginning, most tourists have gone on whale watching trips from Húsavík, which has justifiably been called "the whale watching capital of Europe". The operation of the Húsavík Whale Centre, Iceland's first information centre dedicated to whales, has further strengthened Húsavík's position.

In recent decades, the Marine Research Institute of Iceland has engaged in research on whales in Icelandic waters. Counts of humpback whales and blue whales conducted from whaling vessels in 1970–1985 revealed an increase in the populations of these species. Among other things, it was estimated that the annual increase in the humpback whale population had been 10–15% during the count, and 5% in the blue whale population.

*Humpback whales are frequently sighted on whale watching tours.*

In 1987, 1989, 1995 and 2001, counts were taken of large whales, dolphins and porpoises in the North Atlantic, but the uncertainty factors of these counts

## Research and Whale Watching

The flukes of humpback whales are like our fingerprints. No two flukes are alike.

*A minke whale eyes its observers.*

were too large for any definitive conclusion on whether the populations of these species had grown or not.

In 1998, the Húsavík Whale Centre laid the foundation for co-operation between the Marine Research Institute and whale watching companies regarding whale research. Hopefully, this co-operation will grow in years to come. The eventual goal is to have one staff who will conduct studies of whale behaviour and take photographs of whales.

Whale research is still in its early stages and scientific knowledge in the field is in many ways incomplete. Research aboard whale watching boats should encourage an even more positive attitude towards whale watching, and at the same time promote scientific knowledge and general understanding of marine life and the coastal environment of Iceland.

We ask everyone possessing good photographs of whales to send copies to:

The Húsavík Whale Centre
P.O. Box 172
640 Húsavík
e-mail: icewhale@centrum.is

Photographs should be accompanied by information on where and when they were taken, as well as the name and address of the sender. All those who send photographs will receive a certificate of acknowledgement from the Húsavík Whale Centre.

**APPROACHING THE WHALES**

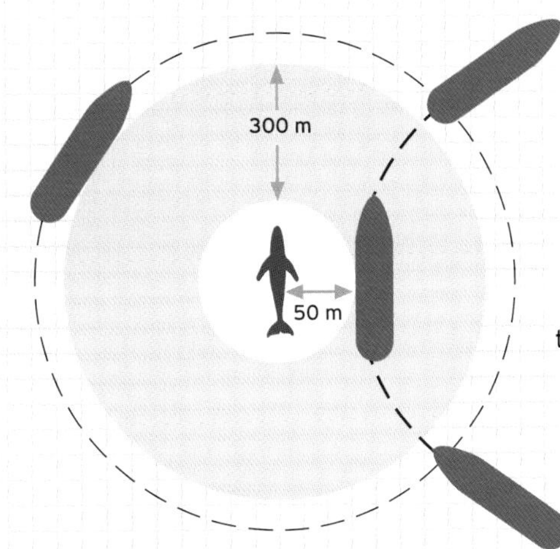

When entering the whale grounds the whales should be approached with caution, from the side, from behind or from the front. Sometimes the whales will approach the boats of their own initiative. If this happens, the propellers of the boat are stopped.

**Whale watching boats never approach whales directly from behind, because then the whales can't see the boat. Passing between cows and their calves should also be avoided.**

# Whale Watching Excursions in Iceland

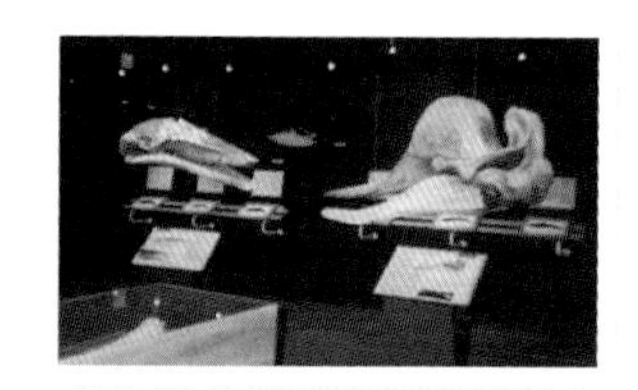

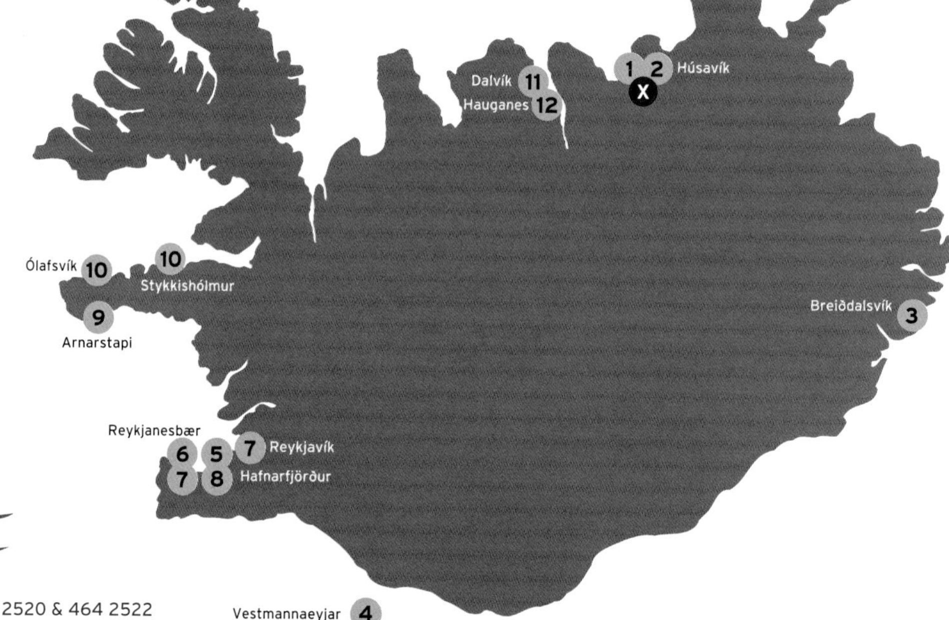



**HVALAMIÐSTÖÐIN Á HÚSAVÍK**
**THE HÚSAVÍK WHALE CENTRE**
The harbour area, Húsavík
☎ 464 2520 & 464 2522
icewhale@centrum.is
www.icewhale.is

X LOCATION ON MAP

Whale Watching and Whale Stranding hotline: 464 2520 & 464 2522

LOCATION ON MAP

**1** **NORÐURSIGLING**
Gamli Baukur, Húsavík
☎ 464-2350
info@nordursigling.is
www.nordursigling.is

**2** **HVALAFERÐIR**
Garðarsbraut 6, Húsavík
☎ 464-2551
www.hvalaferdir.is

**3** **FERÐAÞJÓNUSTAN ÁKI**
760 Breiðdalsvík
☎ 864-0246

**4** **VÍKING BÁTAFERÐIR**
Vestmannaeyjar
☎ 852-7652 & 896-8986
viking@boattours.is
www.boattours.is

**5** **ELDING**
Hafnarfjörður
☎ 555-3565 & 692-4210
elding@islandia.is
www.islandia.is/elding

**6** **HÖFRUNGA- OG HVALASKOÐUN**
Reykjanesbær
☎ 421-7777 & 896-5598
hring@ismennt.is
www.arctic.ic/itn/whale

LOCATION ON MAP

**7** **HVALSTÖÐIN**
Reykjavík/Reykjanesbær
☎ 421-2660 & 895-2523
gestur@marine-marvels.com
www.whalewatching.is

**8** **HÚNI**
Hafnarfjörður
☎ 894-1388, fax: 555-2758
huni@islandia.is
www.islandia.is/huni

**9** **BÁTSFERÐIR ARNARSTAPA/ SNJÓFELL**
Snæfellsnes
☎ 435-6783 & 894-2832
snjofell@snjofell.is

**10** **SÆFERÐIR**
Stykkishólmur/Ólafsvík
☎ 438-1450 & 438-1350
saeferdir@saeferdir.is
www.saeferdir.is

**11** **SJÓFERÐIR**
Dalvík
☎ 466-3355 & 892-3658
863-2555
sjoferdir@simnet.is

**12** **NÍELS JÓNSSON**
Hauganes
☎ 466-1690 & 852-2606

# Birds

*Although those majestic marine mammals, the whales, are naturally the main attraction of whale watching trips, there are many other sights worth seeing. For instance, it may be worthwhile to keep an eye out for sea birds. A flock of birds often indicates schools of fish – and thereby whales. Of the many seabirds seen on whale watching excursions around Iceland these are probably the most common.*

**The northern fulmar** is one of the most common nesting birds in Iceland. It is often confused with various species of seagull, such as the kittiwake, but it has several unmistakable identifying features, including a rather stocky outline and prominent nostrils. When threatened, the fulmar will regurgitate a foul, oily liquid which it spews at perceived adversaries.

**Kittiwakes** are among the most common sea birds of the North Atlantic, with breeding grounds all along the coast of Iceland. The species is easily distinguished from other seagulls by its black wingtips, dark feet and bright yellow beak. Kittiwakes have three toes only instead of four like other seagulls.

**Great black-backed gull** is the largest species of gull found in Iceland. It is easily recognised by the pitch-black back. These are aggressive and intrepid birds and often wreak havoc in the nesting sites of other birds, such as eider ducks.

**Gannet** is the largest Icelandic seabird, with a wingspan of 180 centimetres. Gannets demonstrate admirable flying skills when feeding, often diving vertically from heights of up to 40 metres.

**Guillemots** belong to the razorbill family, but are easily distinguished from razorbills proper by their thinner, sharper beaks and slim necks. In spring, they congregate by the hundreds of thousands in bird cliffs. For centuries, Icelanders have lowered themselves down cliffs to pick eggs, and many people regard guillemot eggs as a rare delicacy.

**Puffin.** Iceland is home to the largest puffin population in the world. An estimated eight to ten million puffins make their homes on Icelandic shores. Characteristic features of these pretty birds are the colourful beaks and "solemn" demeanour. It is no doubt this second characteristic that earned the puffin its nickname "prófastur" (parson). They appear particularly clumsy when taking flight from the sea, and have to run headlong for huge distances on the surface of the water before finally taking off.

**Terns** are extremely agile flyers and travel the longest distances of all migratory birds. Twice each year they travel a distance of some 18,000 kilometres between their nesting areas in the north and their wintering grounds on the shores of the Antarctic. Terns lay their eggs in vast nesting areas which they defend vigorously.

**The eider** is the only Icelandic duck that lays its eggs almost exclusively by the sea; in fact, eiders are a rare sight inland. Many farmers supplement their income by gathering and cleaning the down of eiders; in return they protect the nesting grounds from all kinds of predators. The arrangement is one of the few true forms of symbiosis between Man and wild animals.

**Cormorants** are mainly found around Breiðafjörður and Faxaflói. They are frequently seen sitting upright on rocks or the edge of cliffs with their wings half extended.

*The greatest threats to the natural environment in the Northern Hemisphere: Oil and chemical pollution.*

*A trip to the beach is an endless source of new and exciting discovery.*

# Seals in Icelandic Waters

*Several species of seal inhabit the Arctic Ocean. Three species of these inquisitive sea mammals are most common in Iceland, although there are occasional sightings of other varieties.*

**Harbour seals** can be one or two metres in length, weighing about 150 kilograms. An estimated 50,000 harbour seals live on the Icelandic coast. Seamen generally regard seals as rather gluttonous competitors. Each harbour seal consumes about two and a half tons of fish per year. Some attempts are therefore made to keep their numbers down by hunting them, but commercial sealing is no longer practised. Harbour seals give birth in spring, and in summer they can often be seen in groups on the coast, basking in the sun.

**Grey seals** are considerably larger than harbour seals, reaching a length of up to three metres and weighing as much as 400 kilograms. Grey seals are not seen nearly as often as harbour seals, and they are more vary of humans than their smaller relatives. Their number has been estimated at five to ten thousand animals. They give birth in the autumn and are not as frequent visitors on shore as harbour seals.

**The hooded seal** is one of the largest Arctic seals. These seals can reach a length of 3.2 metres and weigh up to 400–600 kilograms. Hooded seals are easily distinguished by their enormous heads and inflatable air pouch, which is located at the front of the head. In winter they stay around the edge of the pack ice, but in spring and autumn they come up to the Icelandic coast. Many Icelandic seamen take a dark view of the hooded seal because it takes bites out of netted fish and devours their livers, a seal's delicacy.

*A blue whale. The world's largest living creature. Often sighted on whale watching excursions from Snæfellsnes.*

*The chances og sighting whales increases with the number of people on the lookout. Always scan the horizon, as the eyes will then pick up most movements on the surface of the sea.*

*Dolphins will often frolic in the bow wave of the boat.*

*Sometimes the whales choose to approach the boats. Here is a gentle minke whale indulging in some "man watching".*

## Enjoy!

We are guests in the marine kingdom when we visit the whales, the birds and the seals in their natural habitats. We should treat them with respect, approach them with care, and savour our moments with them. Whales are peaceful and often inquisitive animals, and will not attack the boat despite their advantage in size. There is no need to suppress your cheers when the humpback whale leaps out of the water next to the boat, when the blue whale glides past the side, or when the minke whale flips over and offers its belly like a purring cat.

*The Authors*

# References:


Carwardine, Mark: Whales, 1995. *Dolphins and porpoises.*
Dorling Kindersley, London.

Deimer, Petra: 1990. *Das Buch der Wale.*
Heyne, München.

*Íslendingar, hafið og auðlindir þess.* 1994. Published by the Science Society of Iceland. The University of Iceland, Reykjavík.

Johannsen, Willy: 1990. *Hvalernes verden.*
Skarv, Copenhagen.

Ridgway, Sam H.: 1972. *Mammals of the Sea. Biology and Medicine.*
Charles C. Thomas, Springfield.

Sigurður Ægisson, Jón Ásgeir í Aðaldal, Jón Baldur Hlíðberg: 1997.
*Icelandic Whales, Past and Present.*
Forlagið, Reykjavík.

Slijper, E. J.: 1979. *Whales.*
Cornell, Ithaca.

Trausti Einarsson: 1987. *Hvalveiðar við Ísland 1600 – 1939.*
The Cultural Fund, Reykjavík.

*Undraveröld dýranna; 1988. Spendýr.*
Fjölvaútgáfan, Reykjavík.

*Various publications of the Iceland Marine Research Institute, 1982,1988 and 1992.* Jóhann Sigurjónsson, Gísli Víkingsson.

**Aknowledgements**
Our deepest gratitude to the many people who assisted us in the preparation of this book, Mark Carwardine, Martin Camm, Erich Hoyt, Jeff Foster, Heimir Harðarson, Friðþjófur Helgason and the staff of The Húsavík Whale Centre.

**PICTURE CREDITS**
**Cover** Heimir Harðarson; **4** Heimir Harðarson; **5** Jón Ásgeir; **6** PhotoDisc Inc.; **11** Heimir Harðarson; **12** Húsavík Whale Centre; **14** Heimir Harðarson; **16** Heimir Harðarson; **17** Friðþjófur Helgason; **19** Heimir Harðarson; **21** Heimir Harðarson; **23** Heimir Harðarson; **27** Heimir Harðarson (centre); Húsavík Whale Centre (right); **29** Heimir Harðarson (centre; right); **33** Unknown; **35** DV Photo Archive; **37** Ocean Futures; **39** Jón Ásgeir; **41** Mark Carwardine; **43** Jón Ásgeir; **47** Friðþjófur Helgason; **49** Heimir Harðarson; **50** Húsavík Whale Centre (left); Heimir Harðarson (right); **51** Heimir Harðarson; **54** Jóhann Óli Hilmarsson; **55** Jóhann Óli Hilmarsson (left; top right); PhotoDisc Inc. (bottom right); **56** Jón Ásgeir; **57** Jóhann Óli Hilmarsson; **58** Heimir Harðarson; **59** Heimir Harðarson; **60** Heimir Harðarson; **61** Heimir Harðarson; **62** Heimir